MW01593318

30 Days of
Less Tech
W O R K B O O K

a guided digital detox adventure in
using less tech and living more life

Pattie Belle Hastings

For my greatest teacher, August Layne Akselsen.

"In a world dominated by screens, stepping offline is a subversive act of self-care and rebellion."

~ Ken Breniman,
Subversive Acts of Humanity: A Survival Guide for Choosing Evolution over Self-Destruction

Contents

2   Preface

4   Introduction

6   How to Use This Book

8   The U.N.P.L.U.G. Framework

11   30 Days of Less Tech

12-71   Daily Detox Pages

72-79   Notes, doodles, collage, mind maps

80   After the 30 Days

82   My Tech Recovery Snapshot

83   My U.N.P.L.U.G Plan

86   My Relapse Plan

87   Boredom Practice Menu

89   How to Work with Me!

Preface

Why I Care So Much About Your Attention

You and I probably have something in common:

We live inside technology. We work with it, create with it, rely on it, and sometimes feel quietly crushed by it.

I'm not writing to you from a cabin in the woods with no Wi-Fi. I'm a designer, a professor of Interactive Media + Design, and a lifelong digital creator. I've spent decades teaching students how to design interfaces, build digital products, experiment with emerging tech, and now, increasingly, work with AI.

I've also spent decades watching what all of that does to our minds, our bodies, and our hearts.

I've seen:

- students who are brilliant and creative, but can't sit through a 10-minute exercise without reaching for their phones

- professionals who are "always on," yet constantly feel behind

- artists and leaders who feel their attention has been chopped into such tiny fragments that deep work and deep rest feel impossible

And I have watched my own relationship with technology swing between fascination, overuse, avoidance, and careful redesign.

My background is a strange mix:

- I've been a successful visual artist and performance artist, making work about cyborg bodies and media-saturated lives.

- I spent a Fulbright year in Oslo doing design research on mobile devices.

- I've taught generations of students how to design the very interfaces and experiences that now compete for our attention.

- I'm steeped in behavior change, meditation practices, 12-Step–informed recovery ideas, and nervous-system work.

In other words: I understand the hooks from the inside, as a maker. I understand the harm from the outside, as a teacher, facilitator, caregiver, and human. And I understand that shaming people for "too much screen time" is useless in a world where the default settings are quietly addictive. That's why I care so fiercely about helping you change your relationship with your devices without:

- demonizing technology

- glorifying grind culture

- or pretending this is just a matter of "better discipline"

I know how persuasive UX and UI patterns are. I know how AI tools can become yet another place to hide from fatigue, loneliness, or overwhelm. I know what it's like to try to meditate and instead feel like your nervous system is on fire. And I know that for many people, tech overuse isn't just a "bad habit," it overlaps with anxiety, trauma, caregiving, grief, and genuine addiction.

So what I can offer you is this:

- A designer's eye for how your apps and feeds are built to keep you hooked

- An educator's ability to break complex patterns into small, doable experiments

- An artist's belief that your creativity, boredom, and weirdness are not problems – they're part of the cure

- A recovery-informed, nervous-system-informed approach that doesn't treat you like a machine to optimize, but a human to support

This planner is one expression of that work. It's 30 days of guided experiments in using less tech and living more life, designed by someone who has spent years inside the attention economy, asking:

"How do we stay human in all of this?"

I can't walk this journey for you. But I can walk it with you, page-by-page, day-by-day, practice-by-practice. If you're willing to bring your honesty, your curiosity, and your very imperfect self to these pages, I'll bring my decades of design, teaching, research, and lived experience to meet you.

Let's see what becomes possible when your attention belongs to you again.

Pattie Belle Hastings

Introduction

Your Brain vs. the Attention Economy
(and Why AI Raises the Stakes)

Let's tell the truth at the start. You did not "fail" at using technology. Technology was carefully engineered to win at using you.

Most modern digital products are built on a simple economic reality:

If they can keep your attention, they can monetize it.

That means billions of dollars of research into:

- What colors and sounds make you tap faster

- How long to delay a notification for maximum impact

- How infinite scroll can keep you in a loop

- How recommendation algorithms can predict what will keep you watching, swiping, or chatting

This is the attention economy. Your time, your gaze, your emotional reactions, your boredom, your loneliness, and your money are the raw materials for an insatiable greed machine.

And now we have AI layered on top of this:

- AI-curated feeds that learn which content keeps you there the longest

- Addictive AI chatbots that can respond in ways that feel emotionally tuned to you

- AI-generated content that can flood every channel with more stuff than any human could ever absorb

The result? Your nervous system is swimming in constant stimulation and micro-dopamine hits from notifications:

- Tiny red badges

- New messages and likes

- The "maybe there's something better if I scroll just a little more" feeling

- AI tools promising faster, easier, more efficient everything, in exchange for more of your attention

Over time, this barrage has consequences:

- Ordinary life starts to feel flat compared to your feeds.

- Sitting still with your own thoughts feels intolerable.

- You reach for your phone without deciding to.

- Your attention becomes fragmented; deep work and deep rest both feel out of reach.

Then, when you try to meditate, create, or even have a quiet meal, you run smack into agitation, boredom, and the itch to check your device. It's easy to conclude:

"I'm broken."

"I have no willpower."

"I just don't have a good attention span."

In reality, your brain and nervous system are doing exactly what they were trained to do in an environment designed to keep them over-activated. This workbook is one way to push back.

What this book is for

This is a 30-day process for:

- Turning down the volume on the most addictive, manipulative parts of your tech use (including AI tools).

- Understanding how your phone, feeds, and apps are designed to grab you, and what that does to your brain.

- Practicing small, specific acts of selective abstinence, not from all technology, but from the hooks that hurt you most.

- Noticing what you're actually feeling and needing before, during, and after tech use.

- Rebuilding your capacity for boredom, stillness, creativity, and connection.

You'll start to notice the purposeful user experience and user interface patterns, attention hijacks, and the way AI systems learn what keeps you engaged. But this is not a tech theory book. It's a practice book. There are plenty of other books on the research and theory. This a is a call to action in book form.

You'll write. You'll track. You'll experiment. Some days you'll feel powerful. Some days you'll feel irritated and messy. All of it counts.

Thirty days will not undo years of conditioning, but it will give you:

- A deeper understanding of the game being played with your attention

- A clearer sense of your own habits, patterns, and triggers

- A handful of embodied habits and boundaries you can carry forward

You don't have to renounce technology. You do have to decide whose interests your daily behavior is serving: the companies that profit from your time, or the human being who lives your life. This workbook is on your side.

How to Use This Book

A 30-day lab for your attention, not a perfection project

There is no one "correct" way to use this workbook, but there are ways to make it more supportive and less punishing. Think of it as setting up a lab for your own attention and nervous system.

Begin with honesty, not heroics

On Day 1, you'll name your detox and set your intentions. Don't write what you think you "should" want. Write what you actually want back from your devices:

"I want my evenings back."

"I want to stop chatting with AI instead of humans."

"I want my brain to feel less fried."

"I want to be able to read a book again."

"I want my kids/partner/students to see my eyes more than they see my screen."

You can start any day of the week. **If you miss a day, do not start over in shame. Just turn the page and keep going.** Recovery and change are not linear.

Give each day a small, protected window

Plan on 10–20 minutes a day for this work.

Each daily spread asks you to:

- Read the FOCUS, ACTION, and WHY

- Choose what you'll abstain from today (an app, a behavior, a time of day)

- Jot down what you're feeling and needing as you move through the day

- Reflect briefly on how it went

- Use the "My Day" section to link your tech choices to actual time and tasks

- Optionally, use the sketch area to visually process anything that comes up

If you only have five minutes, prioritize:

Action

Abstinence

Feeling/Needing

Everything else is a bonus.

Understand that discomfort is part of the process

At some point you will:

- Get bored

- Feel restless or twitchy

- Want to check your phone "just this once"

- Feel silly writing or drawing in a workbook

- Wonder if anything is "working"

This is not a sign to quit. It is exactly the terrain you're here to explore. When you stay with that discomfort for a few moments longer than usual (without reaching for a device) you're training your nervous system:

> *"We can feel this and not run away. We can survive a quiet moment."*

This is especially true for practices that look simple on paper, like:

- The Morning Delay Rule

- Boredom experiments

- Tech-free meals or commutes

- Photo fasts

- Short "dopamine fasting" windows from high-stimulation apps or AI tools

The first impact is often irritation. The second impact, if you stay, is insight.

Adapt the intensity to your life

There are a few ways to move through these 30 days of experiments:

Deep dive: One practice every day for 30 days. Strong medicine.

Gentle arc: 3-4 days a week over two or three months. More spacious.

Layered support: Use this planner while also working with a therapist, coach, sponsor, or support group around tech addiction or related issues.

If you're living with chronic illness, caregiving, trauma, ADHD, or other challenges, you are absolutely welcome to modify practices:

- Shorten time windows

- Soften abstinence goals ("less" instead of "none")

- Spread the 30 days out

- Combine two easier days into one

The point is not heroism. It's your relationship with tech, with yourself, with your own limits and strengths. Self-compassion and kindness always come first.

Make the pages yours

Some suggestions to have fun with this:

- Use colors: markers, pens, stickers, washi tape ~ whatever makes the book feel alive.

- Collage or tape in printed photos that document your journey.

- Use the SKETCH/DOODLE area for mind maps, comics, diagrams, or visual notes.

- Write about things you would never post publicly. This workbook is a private lab, not a performance.

On Day 28-30 you'll look back through these pages. They will tell a story. It may not be a neat one, but it will be an honest one. That story is the real result.

The U.N.P.L.U.G. Framework

A map for understanding hooks,redesigning habits, and sustaining change

While not overt, this workbook follows the arc of my U.N.P.L.U.G. framework. You don't need to memorize it, but knowing the shape of the journey can help you understand what's happening as you move through the 30 days.

U: Understand Hooks

Learn how the system is designed to grab you

Before we talk about your "habits," we have to talk about the hooks built into your tools. Modern digital products (including AI-powered ones) use specific user experience (UX) and user interface (UI) patterns to keep you engaged:

Infinite scroll: No stopping points, no page breaks. Your brain never gets to a choice point or the "we're done" signal.

Variable rewards: Sometimes the notification is exciting, sometimes it's nothing. This "maybe this time" pattern is the same mechanism used in slot machines.

Red badges and urgency cues: Bright colors, numbers, and "while you were away…" messages spike your threat system and your curiosity.

Autoplay and endless feeds: The next video, the next post, the next recommendation is already there before you decide.

AI-driven personalization: Algorithms and AI models learn what makes you pause, react, argue, or binge, and feed you more of it. That's true whether it's a content feed, a chatbot, or an "assistant" that subtly nudges how you work, shop, or socialize.

Frictionless prompts: "People you may know," "Suggested for you," "Ask me anything…," these are designed to keep you in the system, not remind you that you can walk away.

Understanding hooks means:

- Seeing that you're not weak; you're responding to powerful, well-funded design decisions.

- Noticing your personal weak spots: the apps, times of day, and situations where these patterns grab you hardest.

- Naming the ways AI amplifies this by generating never-ending personalized content, simulating companionship, or smoothing friction so you never reach the point of conscious choice.

In this workbook, the early days (like Screen Time Snapshot, Mood Map, Notification Purge) and many of your reflections are really about seeing the hooks clearly. Once you can say, "Oh, this isn't just me, this is a designed pattern," shame loosens its grip and choice becomes possible.

N: Notice Patterns

Track what you feel, need, and do over and over

Once you understand the hooks, the next step is to notice your patterns:

- When do you reach for your phone or AI tool without thinking?

- What are you feeling right before you do it: bored, lonely, anxious, overwhelmed, curious?

- What need are you trying to meet: connection, escape, control, stimulation, comfort?

- What happens in your body: tight chest, buzzing thoughts, numbness?

The FEELING and NEEDING lines, the boredom practices, and tech-free walks are all about growing this kind of awareness. You're not just tracking "screen time"; you're tracking inner time.

Over these 30 days, you'll start to see your recurring loops:

> *Trigger > Feeling > Device > Short-term relief > Longer-term cost*

Noticing these patterns is not passive. It changes your relationship with them. Once you can say, "Ah, this is my 9 pm loneliness scroll," you're already more conscious than the algorithm expects you to be.

P: Plan Boundaries

Make different choices easier and old habits harder

Willpower is fragile, especially when you're tired, stressed, or dysregulated. That's why you need planned boundaries, not just good intentions.

In this workbook, boundaries show up as:

- Specific rules like the Morning Delay Rule or Screen-Free Hour Before Bed

- Environmental changes like Phone Home or creating Tech-Free Spaces

- Structural shifts like Inbox Intervals and Notification Purges

A good boundary:

- Is clear

- Lives in the world (not just in your head)

- Reduces the number of decisions you have to make in the heat of the moment

You'll experiment with different boundaries throughout these 30 days, and toward the end you'll design your own sustainable ones.

L: Let Go of FOMO

Question what you're afraid of missing and what you're actually missing

Fear of missing out shows up as:

> *"What if there's an emergency?"*
>
> *"What if I fall behind?"*
>
> *"What if I miss a trend, an opportunity, or a message?"*

The Digital Sabbath, Photo Fast, and various abstinence practices are all invitations to sit with FOMO and see what's underneath:

- Sometimes it's a legitimate concern.

- Sometimes it's a social comparison.

- Sometimes it's simply a nervous system trained to equate "offline" with "unsafe" or "irrelevant."

Letting go (even slightly) can reveal something important:

> *Every time you say yes to a screen, you're saying no to something else.*

As you practice being offline (on purpose) you'll start to notice what you've been missing in your offline life: rest, eye contact, silence, daydreaming, creativity, presence.

U: Use Offline Alternatives

Don't just remove the "hit." Replace it.

You can't tear out a compulsive behavior and leave a void. The nervous system hates a void. It will drag you back to whatever gave the fastest relief.

So alongside abstinence, this workbook nudges you to discover:

- Movement instead of scrolling

- Creative bursts instead of doom loops

- Analog mornings instead of nervous system dysregulation

- Slow-media meals instead of background noise

- Real conversation instead of endless "chatting" with AI or strangers online

These alternatives aren't punishment. They're experiments in pleasure and nourishment that don't require an algorithm.

G: Generate Support

Don't do this alone

You can absolutely work through these exercises privately. But know that sustainable change often needs community:

- A person you can text when you want to quit

- A friend or partner to do some of the practices with you

- A therapist, coach, sponsor, or recovery group if your tech use is wrapped up in deeper trauma or addiction

- Communities (online or offline) that share your values around attention, embodiment, and creativity

Sharing your story (Day 27), talking about your patterns, and letting people in on your commitments are powerful forms of accountability.

This book is one layer of support. It remembers what you're trying to do on the days you forget.

OK. Let's UNPLUG

You don't have to consciously manage each letter of U.N.P.L.U.G. as you move through the pages. But as you stay with the process, you'll gradually:

Understand the hooks

Notice your patterns

Plan and live your boundaries

Let go of some FOMO

Use richer offline alternatives

Generate the support you need

That's the real transformation: not just 30 days of "good behavior," but a deeper, kinder, more informed relationship with technology and with yourself.

30 Days of Less Tech

Let's begin the experiments

This section is your daily lab for using less tech and living more life.

Each day you'll:

- Try one small experiment with your tech use
- Notice what you're feeling and needing
- Track how you actually spend your time
- Capture anything that wants to be written, sketched, or mapped

You don't have to be perfect. You just have to keep coming back to the page.

Set Your Intention

Before you start Day 1, take a moment to name why you're here.

Start date:

Right now, my relationship with tech feels:

Over the next 30 days, I want less:

And I want more:

My 30-day intention:

Over the next 30 days, I am choosing to…

You can come back and add to this at any point. The intention is allowed to evolve as you do.

LOCATION

SKETCH | DOODLE | COLLAGE | PHOTO

FEELING (anxious, calm, curious, etc.)

NEEDING (belonging, rest, reassurance, etc.)

TODAY'S CHALLENGE

Day 1: Name Your Detox

Focus: Identity & intention

Action: Give your detox a fun or meaningful name: Project Spacious Mind, The Great Unplug Adventure, Tech-Less Treasure Quest. Write it where you'll see it every day (planner cover, sticky note, phone lock screen).

Why: Your brain latches onto named stories more easily than vague goals, which helps your nervous system feel safer with change. A clear name creates a tiny but powerful identity shift: this isn't just something you're "trying," it's who you're becoming. That identity anchor will make it easier to stick with your choices when cravings or fatigue show up, and it turns the process into an adventure rather than a punishment.

Reflect: Who do you want to become through this detox, and what does your chosen name say about the needs you're honoring?

Name Your Challenge:

TODAY I ABSTAINED FROM:

HOW IT WENT

WHICH ALLOWED ME TO:

MY DAY

TODAY'S FOCUS

TIME

7:00	
7:30	
8:00	
8:30	
9:00	
9:30	

PRIORITY TASKS

10:00	
10:30	
11:00	
11:30	
12:00	
12:30	

CHORES

1:00	
1:30	
2:00	
2:30	
3:00	
3:30	
4:00	
4:30	
5:00	

SPIRIT

5:30	
6:00	
6:30	
7:00	
7:30	

BODY

8:00	
8:30	
9:00	
9:30	

GRATEFUL FOR

INTENTION FOR TOMORROW

DAILY TECH DETOX

LOCATION

FEELING (anxious, calm, curious, etc.)

NEEDING (belonging, rest, reassurance, etc.)

TODAY'S CHALLENGE

Day 2: The Morning Delay Rule

Focus: Awareness before autopilot

Action: Wait at least 15 minutes after waking before touching your phone. Use that time to stretch, drink water, look out a window, breathe, or simply notice the sensations of waking up.

Why: When you avoid starting the day with a flood of notifications, your brain gets a chance to regulate cortisol and settle before taking in stimulation. This builds your capacity to pause before reaching for tech automatically, instead of letting your device set your emotional tone for the day. Over time, this simple delay can reduce morning anxiety, increase clarity, and help you feel more like the author of your day rather than a character in everyone else's feed.

Reflect: What emotional tone did you wake up with today, and how did delaying your phone support the kind of morning you actually needed?

HOW IT WENT

SKETCH | DOODLE | COLLAGE | PHOTO

TODAY I ABSTAINED FROM:

WHICH ALLOWED ME TO:

MY DAY

TODAY'S FOCUS

TIME

Time	
7:00	
7:30	
8:00	
8:30	
9:00	
9:30	
10:00	
10:30	
11:00	
11:30	
12:00	
12:30	
1:00	
1:30	
2:00	
2:30	
3:00	
3:30	
4:00	
4:30	
5:00	
5:30	
6:00	
6:30	
7:00	
7:30	
8:00	
8:30	
9:00	
9:30	

PRIORITY TASKS

CHORES

SPIRIT

BODY

GRATEFUL FOR

INTENTION FOR TOMORROW

LOCATION

SKETCH | DOODLE | COLLAGE | PHOTO

FEELING (anxious, calm, curious, etc.)

NEEDING (belonging, rest, reassurance, etc.)

TODAY'S CHALLENGE

Day 3: Screen Time Snapshot

Focus: Seeing the truth in data

Action: Open your phone's Screen Time or Digital Wellbeing feature and note:
- Top 3 apps by time
- Total pickups
- Average daily use

Write these numbers down without trying to "fix" anything yet.

Why: Looking at your actual usage gives your brain concrete feedback instead of vague guilt, which can calm the shame response in your nervous system. This builds your capacity to make tech decisions based on reality rather than stories like "I'm on my phone all the time" or "It's not that bad." With data in front of you, you can choose one small area to change and free up energy that's been leaking into unconscious habits.

Reflect: Which number hit you the hardest emotionally, and what does that feeling tell you about the life you're craving underneath the habit?

TODAY I ABSTAINED FROM:

HOW IT WENT

WHICH ALLOWED ME TO:

MY DAY

DATE:

TODAY'S FOCUS

PRIORITY TASKS

- ⚬
- ⚬
- ⚬
- ⚬
- ⚬

CHORES

- ⚬
- ⚬
- ⚬
- ⚬
- ⚬

SPIRIT

BODY

GRATEFUL FOR

TIME

Time	
7:00	
7:30	
8:00	
8:30	
9:00	
9:30	
10:00	
10:30	
11:00	
11:30	
12:00	
12:30	
1:00	
1:30	
2:00	
2:30	
3:00	
3:30	
4:00	
4:30	
5:00	
5:30	
6:00	
6:30	
7:00	
7:30	
8:00	
8:30	
9:00	
9:30	

INTENTION FOR TOMORROW

LOCATION

SKETCH | DOODLE | COLLAGE | PHOTO

FEELING (anxious, calm, curious, etc.)

NEEDING (belonging, rest, reassurance, etc.)

TODAY'S CHALLENGE

Day 4: Mood Map

Focus: The emotion–action link

Action: For one day, each time you reach for your phone, pause and silently ask: "What am I feeling right now?" (e.g., bored, lonely, restless, anxious, excited). Jot down a single feeling word each time.

Why: When you link phone use to specific emotional states, your brain starts to see the pattern instead of treating each swipe as a separate event, which reduces that foggy, "Where did my time go?" sensation. This awareness builds your capacity to choose other ways to meet the underlying need instead of automatically using tech as emotional anesthesia. Over time, this can reduce emotional whiplash, increase self-compassion, and give you clearer signals about what your heart and body actually need.

Reflect: Which feeling showed up most often before you grabbed your device, and what deeper need was that feeling pointing toward (comfort, connection, relief, stimulation, belonging)?

TODAY I ABSTAINED FROM:

HOW IT WENT

WHICH ALLOWED ME TO:

MY DAY

TODAY'S FOCUS

TIME

Time	
7:00	
7:30	
8:00	
8:30	
9:00	
9:30	
10:00	
10:30	
11:00	
11:30	
12:00	
12:30	
1:00	
1:30	
2:00	
2:30	
3:00	
3:30	
4:00	
4:30	
5:00	
5:30	
6:00	
6:30	
7:00	
7:30	
8:00	
8:30	
9:00	
9:30	

PRIORITY TASKS

CHORES

SPIRIT

BODY

GRATEFUL FOR

INTENTION FOR TOMORROW

DAILY TECH DETOX

LOCATION

FEELING (anxious, calm, curious, etc.)

NEEDING (belonging, rest, reassurance, etc.)

TODAY'S CHALLENGE

Day 5: Single-Tasking

Focus: Reclaim attention span

Action: Choose one activity today (like emailing, cooking, cleaning, reading, or writing) and do it start-to-finish without checking your phone or multitasking. Set a timer if it helps and commit to staying with just that one thing.

Why: Sustained focus allows your brain to complete the full attention cycle, which calms the nervous system and reduces the constant micro-stress of switching tasks. Practicing single-tasking builds your capacity to resist the urge to "just check" during important work or conversations. Over time, this strengthens your confidence, helps you finish more of what you start, and creates a felt sense of competence that scrolling can never give you.

Reflect: How did it feel in your body and identity to complete something without interruption—more scattered, more steady, or more powerful?

HOW IT WENT

SKETCH | DOODLE | COLLAGE | PHOTO

TODAY I ABSTAINED FROM:

WHICH ALLOWED ME TO:

MY DAY

DATE:

TODAY'S FOCUS

TIME	
7:00	
7:30	
8:00	
8:30	
9:00	
9:30	

PRIORITY TASKS

-
-
-
-
-

10:00	
10:30	
11:00	
11:30	
12:00	
12:30	
1:00	
1:30	

CHORES

-
-
-
-
-

2:00	
2:30	
3:00	
3:30	
4:00	
4:30	
5:00	
5:30	

SPIRIT

6:00	
6:30	
7:00	
7:30	

BODY

8:00	
8:30	
9:00	
9:30	

GRATEFUL FOR

INTENTION FOR TOMORROW

LOCATION

FEELING (anxious, calm, curious, etc.)

NEEDING (belonging, rest, reassurance, etc.)

TODAY'S CHALLENGE

Day 6: Gray-Scale Mode Day

Focus: See the trick

Action: Turn your phone display to grayscale for the day (Settings > Accessibility > Display & Text Size). Leave it that way for as long as you can and notice how your urges change.

Why: Bright colors are designed to spike your dopamine and keep your brain chasing tiny rewards, which keeps your nervous system in a low-level state of alert. Removing color dampens that reward loop and reveals how much of your behavior is driven by design—not by real desire. This experiment builds your capacity to see through manipulative interfaces, reduces compulsive checking, and can make your real-world surroundings feel more vivid by contrast.

Reflect: When your apps lost their color, what did you notice about your cravings, and what does that reveal about the kind of stimulation your nervous system truly needs?

SKETCH | DOODLE | COLLAGE | PHOTO

TODAY I ABSTAINED FROM:

HOW IT WENT

WHICH ALLOWED ME TO:

MY DAY

DATE:

TODAY'S FOCUS

PRIORITY TASKS

-
-
-
-
-

CHORES

-
-
-
-
-

SPIRIT

BODY

GRATEFUL FOR

TIME

Time	
7:00	
7:30	
8:00	
8:30	
9:00	
9:30	
10:00	
10:30	
11:00	
11:30	
12:00	
12:30	
1:00	
1:30	
2:00	
2:30	
3:00	
3:30	
4:00	
4:30	
5:00	
5:30	
6:00	
6:30	
7:00	
7:30	
8:00	
8:30	
9:00	
9:30	

INTENTION FOR TOMORROW

DAILY TECH DETOX

LOCATION

FEELING (anxious, calm, curious, etc.)

NEEDING (belonging, rest, reassurance, etc.)

TODAY'S CHALLENGE

Day 7: Evening Audit

Focus: Reflection & celebration

Action: Before bed, take a few minutes and answer these questions:
- When today did I feel most present?
- What moment of the week felt most like "freedom from the feed"?

Capture these in brief phrases, not essays.

Why: Reviewing your day and week helps your brain consolidate positive changes, teaching your nervous system that shifting your tech use is safe and rewarding. This builds your capacity to notice small wins instead of only tracking failures or slip-ups. Over time, an evening audit turns into a ritual of self-respect, reduces shame, and strengthens your sense of yourself as someone who can choose differently.

Reflect: As you look back on this week, who are you becoming in your relationship with technology, and how does your body respond to that truth?

HOW IT WENT

SKETCH | DOODLE | COLLAGE | PHOTO

TODAY I ABSTAINED FROM:

WHICH ALLOWED ME TO:

MY DAY

TODAY'S FOCUS

TIME

7:00	
7:30	
8:00	
8:30	
9:00	
9:30	

PRIORITY TASKS

10:00
10:30
11:00
11:30
12:00
12:30
1:00
1:30

CHORES

2:00
2:30
3:00
3:30
4:00
4:30
5:00

SPIRIT

5:30
6:00
6:30
7:00
7:30

BODY

8:00
8:30
9:00
9:30

GRATEFUL FOR

INTENTION FOR TOMORROW

LOCATION

FEELING (anxious, calm, curious, etc.)

NEEDING (belonging, rest, reassurance, etc.)

TODAY'S CHALLENGE

Day 8: Tech-Free Dining

Focus: Presence with people or self

Action: Keep all devices off the table for every meal today, whether you're alone or with others. If you eat alone, simply eat; if you're with people, keep phones out of reach and off the table.

Why: Eating without screens helps your brain shift out of "input mode" and into sensory awareness, which supports digestion and nervous system regulation. This builds your capacity to be present with food and people instead of treating meals as a backdrop for more content. Over time, tech-free meals can deepen connection, reduce mindless snacking, and turn eating into a daily ritual of care rather than another place you multitask.

Reflect: What emotional needs (connection, comfort, rest, pleasure) were better met when your meal wasn't competing with a screen?

SKETCH | DOODLE | COLLAGE | PHOTO

TODAY I ABSTAINED FROM:

HOW IT WENT

WHICH ALLOWED ME TO:

MY DAY

DATE:

TODAY'S FOCUS

PRIORITY TASKS

- ⚬
- ⚬
- ⚬
- ⚬
- ⚬

CHORES

- ⚬
- ⚬
- ⚬
- ⚬
- ⚬

SPIRIT

BODY

GRATEFUL FOR

TIME

| 7:00 |
| 7:30 |
| 8:00 |
| 8:30 |
| 9:00 |
| 9:30 |
| 10:00 |
| 10:30 |
| 11:00 |
| 11:30 |
| 12:00 |
| 12:30 |
| 1:00 |
| 1:30 |
| 2:00 |
| 2:30 |
| 3:00 |
| 3:30 |
| 4:00 |
| 4:30 |
| 5:00 |
| 5:30 |
| 6:00 |
| 6:30 |
| 7:00 |
| 7:30 |
| 8:00 |
| 8:30 |
| 9:00 |
| 9:30 |

INTENTION FOR TOMORROW

LOCATION

SKETCH | DOODLE | COLLAGE | PHOTO

FEELING (anxious, calm, curious, etc.)

NEEDING (belonging, rest, reassurance, etc.)

TODAY'S CHALLENGE

Day 9: Inbox Interval

Focus: Reduce reactive checking

Action: Choose three specific windows today to check messages (for example, 9 am, 1 pm, and 5 pm) and commit to avoiding email and DMs outside those times. Write the times down and treat them like appointments.

Why: Batching messages helps your brain shift from constant vigilance to focused problem-solving, reducing the stress hormones that come from perpetual "what if I missed something?" checking. This builds your capacity to set boundaries around responsiveness rather than living in a state of emotional interruption. In the long run, this can protect your focus at work, lower anxiety, and reinforce an identity where your worth isn't measured by how instantly you reply.

Reflect: How did limiting message checks impact your sense of control, and what need are you really trying to meet when you usually refresh your inbox over and over (security, approval, predictability)?

TODAY I ABSTAINED FROM:

HOW IT WENT

WHICH ALLOWED ME TO:

MY DAY

TODAY'S FOCUS

TIME	
7:00	
7:30	
8:00	
8:30	
9:00	
9:30	

PRIORITY TASKS

-
-
-
-
-

10:00	
10:30	
11:00	
11:30	
12:00	
12:30	
1:00	
1:30	

CHORES

-
-
-
-
-

2:00	
2:30	
3:00	
3:30	
4:00	
4:30	
5:00	

SPIRIT

5:30	
6:00	
6:30	
7:00	
7:30	

BODY

8:00	
8:30	
9:00	
9:30	

GRATEFUL FOR

INTENTION FOR TOMORROW

LOCATION

SKETCH | DOODLE | COLLAGE | PHOTO

FEELING (anxious, calm, curious, etc.)

NEEDING (belonging, rest, reassurance, etc.)

TODAY'S CHALLENGE

Day 10: Notification Purge

Focus: Silence the pings

Action: Turn off all non-essential notifications for 24 hours (social media, news, shopping apps, most email alerts). Keep only what you genuinely need for safety or true responsibilities.

Why: Every ping jolts your brain's threat-and-reward systems, keeping your nervous system on edge even when nothing urgent is happening. Silencing these alerts builds your capacity to return your attention to what you care about instead of letting apps yank you around by the collar. Over time, fewer notifications can mean fewer stress spikes, more deep work, and a calmer sense of living your life from the inside out rather than reacting to every digital nudge.

Reflect: Which notifications were hardest to silence emotionally, and what part of you believed you needed them to feel safe, seen, or in control?

TODAY I ABSTAINED FROM:

HOW IT WENT

WHICH ALLOWED ME TO:

MY DAY

TODAY'S FOCUS

TIME

	7:00
	7:30
	8:00
	8:30
	9:00
	9:30

PRIORITY TASKS

-
-
-
-
-

	10:00
	10:30
	11:00
	11:30
	12:00
	12:30
	1:00
	1:30

CHORES

-
-
-
-
-

	2:00
	2:30
	3:00
	3:30
	4:00
	4:30
	5:00
	5:30

SPIRIT

	6:00
	6:30
	7:00
	7:30

BODY

	8:00
	8:30
	9:00
	9:30

GRATEFUL FOR

INTENTION FOR TOMORROW

LOCATION

FEELING (anxious, calm, curious, etc.)

NEEDING (belonging, rest, reassurance, etc.)

TODAY'S CHALLENGE

Day 11: The Boredom Experiment

Focus: Befriend stillness

Action: When boredom arises today, resist the urge to fill it with a screen. Instead, pause, feel the sensations in your body (restlessness, heaviness, itchiness), and simply breathe through them for a minute or two.

Why: Staying with boredom teaches your brain that a lack of stimulation isn't dangerous, which calms the nervous system's urge to grab quick dopamine from your device. This builds your capacity to hold discomfort long enough for genuine curiosity and creativity to emerge. Over time, befriending boredom can unlock new ideas, deepen self-trust, and reveal what you truly want to do when you're not numbing out.

Reflect: When you didn't immediately escape boredom with a screen, what deeper need or desire began to surface underneath the restlessness?

HOW IT WENT

SKETCH | DOODLE | COLLAGE | PHOTO

TODAY I ABSTAINED FROM:

WHICH ALLOWED ME TO:

MY DAY

DATE:

TODAY'S FOCUS

TIME

Time	
7:00	
7:30	
8:00	
8:30	
9:00	
9:30	
10:00	
10:30	
11:00	
11:30	
12:00	
12:30	
1:00	
1:30	
2:00	
2:30	
3:00	
3:30	
4:00	
4:30	
5:00	
5:30	
6:00	
6:30	
7:00	
7:30	
8:00	
8:30	
9:00	
9:30	

PRIORITY TASKS

CHORES

SPIRIT

BODY

GRATEFUL FOR

INTENTION FOR TOMORROW

DAILY TECH DETOX

LOCATION

SKETCH | DOODLE | COLLAGE | PHOTO

FEELING (anxious, calm, curious, etc.)

NEEDING (belonging, rest, reassurance, etc.)

TODAY'S CHALLENGE

Day 12: Tech-Free Commute or Walk

Focus: Sensory reconnection

Action: Take at least one commute, walk, or movement block today without headphones, podcasts, or screens. Let your eyes wander, notice sounds, smells, and the feeling of your body moving through space.

Why: Unplugged movement gives your brain "drift time," which helps process emotions, consolidate memories, and reset your over-stimulated nervous system. This builds your capacity to tolerate being alone with your thoughts without needing constant digital company. Over time, tech-free walks can become a source of insight, calm, and quiet joy instead of a gap you rush to fill with content.

Reflect: What did you notice (outside or inside) that you rarely give yourself time to feel, and what need was being met by simply being present in motion?

TODAY I ABSTAINED FROM:

HOW IT WENT

WHICH ALLOWED ME TO:

MY DAY

TODAY'S FOCUS

TIME

7:00	
7:30	
8:00	
8:30	
9:00	
9:30	

PRIORITY TASKS

-
-
-
-
-

10:00	
10:30	
11:00	
11:30	
12:00	
12:30	
1:00	
1:30	

CHORES

-
-
-
-
-

2:00	
2:30	
3:00	
3:30	
4:00	
4:30	
5:00	
5:30	

SPIRIT

6:00	
6:30	
7:00	
7:30	

BODY

8:00	
8:30	
9:00	
9:30	

GRATEFUL FOR

INTENTION FOR TOMORROW

LOCATION

SKETCH | DOODLE | COLLAGE | PHOTO

FEELING (anxious, calm, curious, etc.)

NEEDING (belonging, rest, reassurance, etc.)

TODAY'S CHALLENGE

Day 13: Phone Home

Focus: Physical boundaries

Action: Choose a single place in your home where your phone "lives" when you're not using it (a basket, shelf, or charging station) and return it there every time you're done. Treat it like putting away a tool.

Why: Creating a home for your phone helps your brain break the "always within reach" loop, which reduces automatic grabbing and soothes the nervous system by making the device less central. This builds your capacity to have intentional on/off time with tech rather than a constant low-level attachment. Over time, this simple boundary can restore pockets of silence, support face-to-face connection, and remind you that your home is for living, not just for charging devices.

Reflect: When your phone had a defined "home," how did your space and sense of self feel different: more spacious, more grounded, or more in charge?

TODAY I ABSTAINED FROM:

HOW IT WENT

WHICH ALLOWED ME TO:

MY DAY

DATE:

TODAY'S FOCUS

TIME	
7:00	
7:30	
8:00	
8:30	
9:00	
9:30	
10:00	
10:30	
11:00	
11:30	
12:00	
12:30	
1:00	
1:30	
2:00	
2:30	
3:00	
3:30	
4:00	
4:30	
5:00	
5:30	
6:00	
6:30	
7:00	
7:30	
8:00	
8:30	
9:00	
9:30	

PRIORITY TASKS

CHORES

SPIRIT

BODY

GRATEFUL FOR

INTENTION FOR TOMORROW

LOCATION

FEELING (anxious, calm, curious, etc.)

NEEDING (belonging, rest, reassurance, etc.)

TODAY'S CHALLENGE

Day 14: Screen-Free Hour Before Bed

Focus: Rest & recovery

Action: Power down all devices one hour before sleep. Use that hour for analog activities: stretching, reading, journaling, gentle tidying, or quiet conversation.

Why: Screen light and stimulation signal your brain to stay alert, delaying melatonin and keeping your nervous system revved when it should be winding down. Choosing a screen-free hour builds your capacity to transition from "doing and consuming" to "resting and restoring" without digital crutches. Over time, this practice can improve sleep, reduce late-night anxiety spirals, and help you wake feeling more like yourself.

Reflect: What did your body and heart seem to need most in that last hour—comfort, closure, connection, or true rest—and how did being offline help you honor that need?

SKETCH | DOODLE | COLLAGE | PHOTO

TODAY I ABSTAINED FROM:

HOW IT WENT

WHICH ALLOWED ME TO:

MY DAY

TODAY'S FOCUS

TIME

7:00	
7:30	
8:00	
8:30	
9:00	
9:30	
10:00	
10:30	
11:00	
11:30	
12:00	
12:30	
1:00	
1:30	
2:00	
2:30	
3:00	
3:30	
4:00	
4:30	
5:00	
5:30	
6:00	
6:30	
7:00	
7:30	
8:00	
8:30	
9:00	
9:30	

PRIORITY TASKS

CHORES

SPIRIT

BODY

GRATEFUL FOR

INTENTION FOR TOMORROW

LOCATION

SKETCH | DOODLE | COLLAGE | PHOTO

FEELING (anxious, calm, curious, etc.)

NEEDING (belonging, rest, reassurance, etc.)

TODAY'S CHALLENGE

Day 15: Analog Mornings

Focus: Start in your own energy

Action: Spend at least the first 30 minutes of your day offline. Journal, stretch, doodle, sip a warm drink, or sit quietly without checking any devices.

Why: Analog mornings let your brain emerge from sleep without a shock of digital input, which keeps your nervous system from jumping straight into vigilance or comparison. This builds your capacity to hear your own thoughts and feelings before everyone else's opinions and demands rush in. Over time, starting analog can stabilize your mood, deepen your intuition, and help you anchor into who you are before you go online.

Reflect: When you began your day in analog mode, what part of you felt most relieved or seen—your body, your creativity, your emotions, or your sense of purpose?

TODAY I ABSTAINED FROM:

HOW IT WENT

WHICH ALLOWED ME TO:

MY DAY

DATE:

TODAY'S FOCUS

TIME

	7:00
	7:30
	8:00
	8:30
	9:00
	9:30

PRIORITY TASKS

-
-
-
-
-

10:00
10:30
11:00
11:30
12:00
12:30
1:00
1:30

CHORES

-
-
-
-
-

2:00
2:30
3:00
3:30
4:00
4:30
5:00
5:30

SPIRIT

6:00
6:30
7:00
7:30

BODY

8:00
8:30
9:00
9:30

GRATEFUL FOR

INTENTION FOR TOMORROW

LOCATION

FEELING (anxious, calm, curious, etc.)

NEEDING (belonging, rest, reassurance, etc.)

TODAY'S CHALLENGE

Day 16: Digital Detox Date

Focus: Relational presence

Action: Spend intentional time with someone (on a walk, over a meal, or in conversation) with phones on airplane mode and out of sight. Let the interaction be fully screen-free.

Why: Face-to-face presence and eye contact cue your brain to release oxytocin, which calms the nervous system and counters the jittery dopamine spikes of social media. This builds your capacity to feel truly connected without needing to document, text, or check reactions online. Over time, tech-free time with others can deepen intimacy, heal loneliness, and remind you that your worth isn't measured in likes but in lived relationships.

Reflect: How did giving and receiving undivided attention change how you felt about yourself and the relationship—more secure, more valued, more at ease?

SKETCH | DOODLE | COLLAGE | PHOTO

TODAY I ABSTAINED FROM:

HOW IT WENT

WHICH ALLOWED ME TO:

MY DAY

TODAY'S FOCUS

TIME

7:00	
7:30	
8:00	
8:30	
9:00	
9:30	

PRIORITY TASKS

-
-
-
-
-

10:00	
10:30	
11:00	
11:30	
12:00	
12:30	
1:00	
1:30	

CHORES

-
-
-
-
-

2:00	
2:30	
3:00	
3:30	
4:00	
4:30	
5:00	
5:30	

SPIRIT

6:00	
6:30	
7:00	
7:30	

BODY

8:00	
8:30	
9:00	
9:30	

GRATEFUL FOR

INTENTION FOR TOMORROW

DAILY TECH DETOX

LOCATION

FEELING (anxious, calm, curious, etc.)

NEEDING (belonging, rest, reassurance, etc.)

TODAY'S CHALLENGE

Day 17: The 5-Minute Creative Burst

Focus: Creation over consumption

Action: Use at least five reclaimed minutes to make something: doodle, write a haiku, sing, dance, arrange objects, or sketch in your planner. It doesn't have to be "good." It just has to be yours.

Why: Creating activates networks in your brain that support flow and problem-solving rather than passive consumption, giving your nervous system a nourishing form of stimulation. This builds your capacity to reach for self-expression instead of endless scrolling when you feel restless or numb. Over time, tiny creative bursts can restore a sense of aliveness, confidence, and joy that algorithms can't deliver.

Reflect: What did you create today, and what part of your identity or inner need did that small act of making honor (play, meaning, beauty, voice)?

SKETCH | DOODLE | COLLAGE | PHOTO

TODAY I ABSTAINED FROM:

HOW IT WENT

WHICH ALLOWED ME TO:

MY DAY

TODAY'S FOCUS

TIME

7:00	
7:30	
8:00	
8:30	
9:00	
9:30	

PRIORITY TASKS

-
-
-
-
-

10:00	
10:30	
11:00	
11:30	
12:00	
12:30	
1:00	
1:30	

CHORES

-
-
-
-
-

2:00	
2:30	
3:00	
3:30	
4:00	
4:30	
5:00	

SPIRIT

5:30	
6:00	
6:30	
7:00	
7:30	

BODY

8:00	
8:30	
9:00	
9:30	

GRATEFUL FOR

INTENTION FOR TOMORROW

LOCATION

FEELING (anxious, calm, curious, etc.)

NEEDING (belonging, rest, reassurance, etc.)

TODAY'S CHALLENGE

Day 18: Move Your Body

Focus: Embodiment

Action: Trade at least 20 minutes of screen time for movement: walking, dancing, stretching, yoga, cleaning, or gardening. Move in a way that feels possible and kind to your body.

Why: Movement helps your brain metabolize stress hormones and reset dopamine sensitivity, allowing your nervous system to discharge tension built up from screen time. This builds your capacity to regulate emotions through your body rather than through digital distraction. Over time, moving instead of scrolling can improve mood, reduce anxiety, and reinforce a self-image of someone who cares for their body, not just their inbox.

Reflect: How did your emotional state and sense of self shift from before to after moving? Did you feel more powerful, more peaceful, more present, or something else?

TODAY I ABSTAINED FROM:

HOW IT WENT

WHICH ALLOWED ME TO:

MY DAY

TODAY'S FOCUS

TIME

	7:00
	7:30
	8:00
	8:30
	9:00
	9:30

PRIORITY TASKS

- ⊙
- ⊙
- ⊙
- ⊙
- ⊙

10:00
10:30
11:00
11:30
12:00
12:30
1:00
1:30

CHORES

- ⊙
- ⊙
- ⊙
- ⊙
- ⊙

2:00
2:30
3:00
3:30
4:00
4:30
5:00
5:30

SPIRIT

6:00
6:30
7:00
7:30

BODY

8:00
8:30
9:00
9:30

GRATEFUL FOR

INTENTION FOR TOMORROW

LOCATION

SKETCH | DOODLE | COLLAGE | PHOTO

FEELING (anxious, calm, curious, etc.)

NEEDING (belonging, rest, reassurance, etc.)

TODAY'S CHALLENGE

Day 19: Slow-Media Meal

Focus: Mindful consumption

Action: Choose one meal or snack today to enjoy without any parallel media – no TV, podcasts, or phone. Eat slowly, paying attention to taste, smell, texture, and how your body feels as you eat.

Why: Eating without media helps your brain reconnect hunger, pleasure, and fullness signals, which calms the nervous system and reduces sensory overload. This builds your capacity to experience satisfaction without needing extra digital stimulation layered on top. Over time, slow-media meals can improve digestion, curb "doom-scroll snacking," and reconnect you with the simple joy of being nourished.

Reflect: What did you discover about your body's needs (hunger, fullness, comfort, or care) when food wasn't competing with a stream of content?

TODAY I ABSTAINED FROM:

HOW IT WENT

WHICH ALLOWED ME TO:

MY DAY

TODAY'S FOCUS

TIME

Time	
7:00	
7:30	
8:00	
8:30	
9:00	
9:30	
10:00	
10:30	
11:00	
11:30	
12:00	
12:30	
1:00	
1:30	
2:00	
2:30	
3:00	
3:30	
4:00	
4:30	
5:00	
5:30	
6:00	
6:30	
7:00	
7:30	
8:00	
8:30	
9:00	
9:30	

PRIORITY TASKS

CHORES

SPIRIT

BODY

GRATEFUL FOR

INTENTION FOR TOMORROW

LOCATION

SKETCH | DOODLE | COLLAGE | PHOTO

FEELING (anxious, calm, curious, etc.)

NEEDING (belonging, rest, reassurance, etc.)

TODAY'S CHALLENGE

Day 20: Digital Sabbath Sampler

Focus: Rest & renewal

Action: Take a half-day off from non-essential screens (especially social media). Let people know you'll be offline, then step away and simply live your life.

Why: Extended breaks give your brain time to downshift from constant stimulation, allowing your nervous system to move out of fight-or-flight and into rest-and-digest. This builds your capacity to trust that the world can go on without your constant digital presence and that you are still safe and connected. Over time, regular mini-sabbaths can reduce burnout, restore curiosity, and remind you what you enjoy when you're not inside a screen.

Reflect: What emotions surfaced during your offline window (relief, anxiety, grief, ease) and what need were those feelings pointing toward (safety, belonging, freedom, rest)?

TODAY I ABSTAINED FROM:

HOW IT WENT

WHICH ALLOWED ME TO:

MY DAY

TODAY'S FOCUS

TIME

7:00	
7:30	
8:00	
8:30	
9:00	
9:30	
10:00	
10:30	
11:00	
11:30	
12:00	
12:30	
1:00	
1:30	
2:00	
2:30	
3:00	
3:30	
4:00	
4:30	
5:00	
5:30	
6:00	
6:30	
7:00	
7:30	
8:00	
8:30	
9:00	
9:30	

PRIORITY TASKS

CHORES

SPIRIT

BODY

GRATEFUL FOR

INTENTION FOR TOMORROW

LOCATION

FEELING (anxious, calm, curious, etc.)

NEEDING (belonging, rest, reassurance, etc.)

TODAY'S CHALLENGE

Day 21: Photo Fast

Focus: Presence over capture

Action: Spend the day noticing beautiful or interesting moments without taking photos. Instead, describe them in words or quick sketches in your planner.

Why: When you don't outsource memory to your camera, your brain engages more fully with the moment, strengthening the neural pathways for presence and recall. This builds your capacity to experience life directly instead of constantly standing outside yourself as a "content creator." Over time, photo fasts can deepen your appreciation for ordinary moments, reduce performance pressure, and reconnect you with the quiet satisfaction of simply being there.

Reflect: When you didn't reach for the camera, how did your sense of self in the moment change from observer, performer, or archivist to participant?

SKETCH | DOODLE | COLLAGE | PHOTO

TODAY I ABSTAINED FROM:

HOW IT WENT

WHICH ALLOWED ME TO:

MY DAY

TODAY'S FOCUS

PRIORITY TASKS
-
-
-
-
-

CHORES
-
-
-
-
-

SPIRIT

BODY

GRATEFUL FOR

TIME

Time	
7:00	
7:30	
8:00	
8:30	
9:00	
9:30	
10:00	
10:30	
11:00	
11:30	
12:00	
12:30	
1:00	
1:30	
2:00	
2:30	
3:00	
3:30	
4:00	
4:30	
5:00	
5:30	
6:00	
6:30	
7:00	
7:30	
8:00	
8:30	
9:00	
9:30	

INTENTION FOR TOMORROW

DAILY TECH DETOX

LOCATION

FEELING (anxious, calm, curious, etc.)

NEEDING (belonging, rest, reassurance, etc.)

TODAY'S CHALLENGE

Day 22: Identify Your Triggers

Focus: Pattern recognition

Action: List at least three situations that most often lead you to mindless tech use (for example: late-night exhaustion, awkward social pauses, waiting in line, post-lunch slump). Be specific.

Why: Naming triggers helps your brain see the cue/behavior loop clearly, which reduces the sense that your nervous system is being hijacked "out of nowhere." This builds your capacity to anticipate vulnerable moments and prepare supportive alternatives instead of falling into default scrolling. Over time, understanding your triggers can reduce self-blame, increase self-kindness, and make your tech choices feel deliberate rather than mysterious.

TODAY I ABSTAINED FROM:

Reflect: Which trigger feels most tender or loaded for you emotionally, and what deeper need is hiding underneath it (comfort, escape, affirmation, numbing, control)?

HOW IT WENT

WHICH ALLOWED ME TO:

MY DAY

TODAY'S FOCUS

TIME

| 7:00 |
| 7:30 |
| 8:00 |
| 8:30 |
| 9:00 |
| 9:30 |
| 10:00 |
| 10:30 |
| 11:00 |
| 11:30 |
| 12:00 |
| 12:30 |
| 1:00 |
| 1:30 |
| 2:00 |
| 2:30 |
| 3:00 |
| 3:30 |
| 4:00 |
| 4:30 |
| 5:00 |
| 5:30 |
| 6:00 |
| 6:30 |
| 7:00 |
| 7:30 |
| 8:00 |
| 8:30 |
| 9:00 |
| 9:30 |

PRIORITY TASKS

CHORES

SPIRIT

BODY

GRATEFUL FOR

INTENTION FOR TOMORROW

DAILY TECH DETOX

LOCATION

SKETCH | DOODLE | COLLAGE | PHOTO

FEELING (anxious, calm, curious, etc.)

NEEDING (belonging, rest, reassurance, etc.)

TODAY'S CHALLENGE

Day 23: Create Your Rules of Use

Focus: Intentional design

Action: Draft 3–5 personal Tech Boundaries rules, such as: "No screens at meals," "I check messages twice a day," or "I'm off social media by 9 pm." Write them somewhere visible.

Why: Clear, self-chosen rules reduce decision fatigue in your brain, which keeps your nervous system from constantly negotiating with cravings and "just one more" impulses. This builds your capacity to be the architect of your digital life rather than a passive user. Over time, living by your own rules can strengthen self-respect, simplify your days, and support the identity of someone who protects their attention as something precious.

Reflect: Which rule feels like a deep exhale of relief, and which one challenges an old identity or fear about who you need to be online?

TODAY I ABSTAINED FROM:

HOW IT WENT

WHICH ALLOWED ME TO:

MY DAY

DATE:

TODAY'S FOCUS

TIME

	7:00
	7:30
	8:00
	8:30
	9:00
	9:30

PRIORITY TASKS

- ⊙
- ⊙
- ⊙
- ⊙
- ⊙

10:00
10:30
11:00
11:30
12:00
12:30
1:00
1:30

CHORES

- ⊙
- ⊙
- ⊙
- ⊙
- ⊙

2:00
2:30
3:00
3:30
4:00
4:30
5:00

SPIRIT

5:30
6:00
6:30
7:00
7:30

BODY

8:00
8:30
9:00
9:30

GRATEFUL FOR

INTENTION FOR TOMORROW

LOCATION

FEELING (anxious, calm, curious, etc.)

NEEDING (belonging, rest, reassurance, etc.)

TODAY'S CHALLENGE

Day 24: Tech-Free Space

Focus: Environment shapes behavior

Action: Choose one physical area (like your bedroom, dining table, or studio) and declare it a no-screen zone. Remove chargers and devices from that space as much as possible.

Why: When your brain associates a space with rest, creativity, or connection instead of constant alerts, your nervous system can relax more quickly in that environment. This builds your capacity to access different states (sleep, focus, intimacy) without competing digital noise. Over time, dedicated tech-free spaces can become sanctuaries that remind you who you are beyond your online roles and responsibilities.

Reflect: What identity or need does this tech-free space protect for you as an artist, sleeper, friend, lover, thinker, or simply a human who deserves peace?

SKETCH | DOODLE | COLLAGE | PHOTO

TODAY I ABSTAINED FROM:

HOW IT WENT

WHICH ALLOWED ME TO:

MY DAY

DATE:

TODAY'S FOCUS

TIME	
7:00	
7:30	
8:00	
8:30	
9:00	
9:30	
10:00	
10:30	
11:00	
11:30	
12:00	
12:30	
1:00	
1:30	
2:00	
2:30	
3:00	
3:30	
4:00	
4:30	
5:00	
5:30	
6:00	
6:30	
7:00	
7:30	
8:00	
8:30	
9:00	
9:30	

PRIORITY TASKS

CHORES

SPIRIT

BODY

GRATEFUL FOR

INTENTION FOR TOMORROW

DAILY TECH DETOX

LOCATION

SKETCH | DOODLE | COLLAGE | PHOTO

FEELING (anxious, calm, curious, etc.)

NEEDING (belonging, rest, reassurance, etc.)

TODAY'S CHALLENGE

Day 25: Digital Declutter

Focus: Simplify to sustain

Action: Delete at least five unused apps, unsubscribe from junk emails, or close out your sea of open tabs. Release what no longer serves you.

Why: Digital clutter keeps your brain in low-level overwhelm, asking it to track far more than your nervous system can comfortably handle. Clearing it builds your capacity to focus on what actually matters instead of constantly scanning cluttered screens. Over time, decluttering your devices can reduce decision fatigue, ease anxiety, and mirror the emotional relief of cleaning a physical room.

Reflect: As you deleted and unsubscribed, what feelings surfaced (loss, relief, fear, satisfaction) and what do those emotions reveal about your attachment to digital "stuff"?

TODAY I ABSTAINED FROM:

HOW IT WENT

WHICH ALLOWED ME TO:

MY DAY

TODAY'S FOCUS

TIME

7:00	
7:30	
8:00	
8:30	
9:00	
9:30	
10:00	
10:30	
11:00	
11:30	
12:00	
12:30	
1:00	
1:30	
2:00	
2:30	
3:00	
3:30	
4:00	
4:30	
5:00	
5:30	
6:00	
6:30	
7:00	
7:30	
8:00	
8:30	
9:00	
9:30	

PRIORITY TASKS

CHORES

SPIRIT

BODY

GRATEFUL FOR

INTENTION FOR TOMORROW

LOCATION

FEELING (anxious, calm, curious, etc.)

NEEDING (belonging, rest, reassurance, etc.)

TODAY'S CHALLENGE

Day 26: Reclaim Time

Focus: Redirect your energy

Action: Take a chunk of time you've freed up (even 20 minutes) and use it for something that genuinely nourishes you: reading, resting, playing, cooking, learning, connecting, or simply staring out a window.

Why: When your brain experiences a rewarding alternative to scrolling, your nervous system starts to associate breaks with restoration instead of numbing. This builds your capacity to choose life-giving activities when you're tired, stressed, or overstimulated rather than falling back into old loops. Over time, reclaiming pockets of time can shift your entire sense of who your days belong to and what kind of life you're building.

Reflect: What did you do with reclaimed time today, and what long-neglected need did that choice finally honor (rest, joy, meaning, learning, connection)?

SKETCH | DOODLE | COLLAGE | PHOTO

TODAY I ABSTAINED FROM:

HOW IT WENT

WHICH ALLOWED ME TO:

MY DAY

TODAY'S FOCUS

TIME

TIME	
7:00	
7:30	
8:00	
8:30	
9:00	
9:30	
10:00	
10:30	
11:00	
11:30	
12:00	
12:30	
1:00	
1:30	
2:00	
2:30	
3:00	
3:30	
4:00	
4:30	
5:00	
5:30	
6:00	
6:30	
7:00	
7:30	
8:00	
8:30	
9:00	
9:30	

PRIORITY TASKS

CHORES

SPIRIT

BODY

GRATEFUL FOR

INTENTION FOR TOMORROW

LOCATION

FEELING (anxious, calm, curious, etc.)

NEEDING (belonging, rest, reassurance, etc.)

TODAY'S CHALLENGE

Day 27: Share Your Story

Focus: Accountability & voice

Action: Tell someone (out loud or in writing) about your biggest shift from these 27 days. Share what you tried, what changed, and what you want to keep.

Why: Speaking your story out loud helps your brain integrate the experience, signaling to your nervous system that this change is real and important. This builds your capacity to claim a new identity in relation to technology ("I'm someone who protects my attention," "I'm in recovery from digital overload"). Over time, sharing your journey can inspire others, deepen your courage, and remind you that you're not alone in wanting a different relationship with screens.

Reflect: When you shared your story, how did it feel in your body and identity – more exposed, more powerful, more seen, or more committed?

TODAY I ABSTAINED FROM:

HOW IT WENT

WHICH ALLOWED ME TO:

MY DAY

DATE:

TODAY'S FOCUS

TIME	
7:00	
7:30	
8:00	
8:30	
9:00	
9:30	

PRIORITY TASKS

-
-
-
-
-

10:00	
10:30	
11:00	
11:30	
12:00	
12:30	
1:00	
1:30	

CHORES

-
-
-
-
-

2:00	
2:30	
3:00	
3:30	
4:00	
4:30	
5:00	

SPIRIT

5:30	
6:00	
6:30	
7:00	
7:30	

BODY

8:00	
8:30	
9:00	
9:30	

GRATEFUL FOR

INTENTION FOR TOMORROW

LOCATION

FEELING (anxious, calm, curious, etc.)

NEEDING (belonging, rest, reassurance, etc.)

SKETCH | DOODLE | COLLAGE | PHOTO

TODAY'S CHALLENGE

Day 28: Reflection Day

Focus: Integrate & witness

Action: Re-read your notes and reflections from Weeks 1–3. Make a short list of what has improved (sleep, mood, focus, creativity, relationships, energy) and where you still feel tender.

Why: Looking back shows your brain the arc of change, reinforcing the neural pathways that support your new habits and soothing a nervous system that may still doubt your progress. This builds your capacity to see yourself as someone who can change, not someone who is "just like this" forever. Over time, honest reflection (including what still hurts) can deepen self-compassion, reduce all-or-nothing thinking, and help you choose sustainable next steps.

Reflect: What identity story about yourself and technology are you ready to retire, and what new story is quietly asking to take its place?

TODAY I ABSTAINED FROM:

HOW IT WENT

WHICH ALLOWED ME TO:

MY DAY

DATE:

TODAY'S FOCUS

TIME

7:00	
7:30	
8:00	
8:30	
9:00	
9:30	

PRIORITY TASKS

- ○
- ○
- ○
- ○
- ○

10:00
10:30
11:00
11:30
12:00
12:30
1:00
1:30
2:00
2:30

CHORES

- ○
- ○
- ○
- ○
- ○

3:00
3:30
4:00
4:30
5:00
5:30
6:00
6:30
7:00
7:30

SPIRIT

BODY

8:00
8:30
9:00
9:30

GRATEFUL FOR

INTENTION FOR TOMORROW

DAILY TECH DETOX

LOCATION

FEELING (anxious, calm, curious, etc.)

NEEDING (belonging, rest, reassurance, etc.)

TODAY'S CHALLENGE

Day 29: Design Your Sustainable Plan

Focus: Future mapping

Action: Choose three practices from this challenge to keep as long-term habits. Write them clearly and place them where you'll see them often (planner, fridge, lock screen, mirror).

Why: Limiting your focus to a few key habits keeps your brain from overwhelm, which helps your nervous system trust that this new way of living is doable. This builds your capacity to maintain change without perfectionism, integrating tech boundaries into your daily rhythm. Over time, a small, clear plan can reshape your lifestyle quietly but profoundly, protecting your attention, energy, and relationships.

Reflect: Looking ahead, who do you want to be six months from now in your relationship with tech, and how do these three habits serve that version of you?

HOW IT WENT

SKETCH | DOODLE | COLLAGE | PHOTO

TODAY I ABSTAINED FROM:

WHICH ALLOWED ME TO:

MY DAY

TODAY'S FOCUS

TIME

7:00	
7:30	
8:00	
8:30	
9:00	
9:30	

PRIORITY TASKS

-
-
-
-
-

10:00	
10:30	
11:00	
11:30	
12:00	
12:30	
1:00	
1:30	
2:00	
2:30	
3:00	
3:30	
4:00	
4:30	

CHORES

-
-
-
-
-

5:00	
5:30	
6:00	
6:30	
7:00	
7:30	

SPIRIT

8:00	
8:30	
9:00	
9:30	

BODY

GRATEFUL FOR

INTENTION FOR TOMORROW

LOCATION

SKETCH | DOODLE | COLLAGE | PHOTO

FEELING (anxious, calm, curious, etc.)

NEEDING (belonging, rest, reassurance, etc.)

TODAY'S CHALLENGE

Day 30: Celebrate

Focus: Honor your transformation

Action: Plan and carry out a celebration or ritual to mark completing this 30-day experiment—no matter how "perfectly" you did it. Choose something that feels like genuine appreciation for your effort.

Why: Celebration teaches your brain that effort and growth are worthy of reward, which helps your nervous system associate change with pleasure rather than only struggle. This builds your capacity to keep evolving without burning out or slipping into self-criticism. Over time, honoring your milestones—even messy ones—supports an identity of someone who shows up, learns, and keeps going.

Reflect: If you spoke directly to the you who started on Day 1, what would you thank them for, and what promise do you want to make to yourself now?

TODAY I ABSTAINED FROM:

HOW IT WENT

WHICH ALLOWED ME TO:

MY DAY

DATE:

TODAY'S FOCUS

TIME

7:00	
7:30	
8:00	
8:30	
9:00	
9:30	

PRIORITY TASKS

- ⚙
- ⚙
- ⚙
- ⚙
- ⚙

10:00	
10:30	
11:00	
11:30	
12:00	
12:30	
1:00	
1:30	

CHORES

- ⚙
- ⚙
- ⚙
- ⚙
- ⚙

2:00	
2:30	
3:00	
3:30	
4:00	
4:30	
5:00	
5:30	

SPIRIT

6:00	
6:30	
7:00	
7:30	

BODY

8:00	
8:30	
9:00	
9:30	

GRATEFUL FOR

INTENTION FOR TOMORROW

Notes, doodles, collage, mind maps

Notes, doodles, collage, mind maps

Notes, doodles, collage, mind maps

Notes, doodles, collage, mind maps

After the 30 Days

Look back before you rush forward

You made it to the end of this 30-day experiment. Before you decide what's next, pause and look at what actually happened. This isn't a test. It's a debrief ~ a way to harvest what you learned from your own nervous system, not just from the pages.

Use these pages to take stock.

What shifted?

- Where do you notice the biggest changes in your daily life? (Sleep, mood, focus, creativity, relationships, energy, body sensations.)

- Did any apps, platforms, or behaviors lose some of their hold on you?

- When you look at your FEELING / NEEDING entries, what themes show up?

What stayed the same?

- Which habits or hooks still feel just as strong? Name them without judgment.

- Are there times of day or specific situations where you still feel completely hijacked?

- What old patterns came back the moment you were stressed, tired, or lonely?

What surprised you?

- Did anything feel easier than you expected?
- Did any practice you "rolled your eyes at" actually help?
- Did you discover any unexpected pleasures in being offline or less stimulated?

What do you want more of?

This might be:

- More real conversation
- More sleep
- More drawing, writing, or making
- More quiet mornings
- More movement
- More deep work blocks

My Tech Recovery Snapshot

From "before" to "right now"

Before: My relationship with tech felt like...

- Words that described it: (e.g., frantic, numbing, necessary, addictive, chaotic, "fine but draining")
- Typical day: What did mornings, evenings, and breaks look like?
- My biggest pain points: (e.g., sleep, anxiety, distraction, shame, lost time, money, relationships)

After 30 Days – My relationship with tech now feels like...

- Words that describe it today: (even if they're mixed "less chaotic," "honestly, still messy")
- What's noticeably better?
- What's still tender or wobbly?

Three wins I want to remember

These do not have to be heroic. "I turned my phone off at night for the first time in years" is a win. "I saw my own pattern clearly" is a win.

1.

2.

3.

My U.N.P.L.U.G Plan

Turning insights into daily reality

You've just walked through all six letters, more or less. Now you'll sketch the version of U.N.P.L.U.G. that belongs to you. You can keep this high-level or very specific.

U: Understand Hooks

Which hooks matter most in your life?
- Apps, platforms, or devices that give you the most trouble
- Manipulative patterns you know target you: (infinite scroll, autoplay, recommendations, notifications, AI chat at night, etc.)
- One sentence that sums it up: "The main hooks that grab me are…"

N: Notice Patterns

What are your key loops?
- Times of day that are most vulnerable
- Feelings that send you straight to a screen: (lonely, restless, numb, angry, overwhelmed, FOMO…)
- Needs hiding underneath those feelings: (connection, comfort, distraction, escape, validation, rest…)
- Write a few of your clearest loops: "When I feel _____, I usually reach for _____ to get _____."

P: Plan Boundaries

Which concrete boundaries do you want to keep?
- Examples: Morning Delay Rule, Screen-Free Hour Before Bed, Phone Home, Inbox Intervals, Tech-Free Spaces, etc.
- List 3-5 boundaries that feel supportive, realistic, and that you can commit to for the next 30-90 days.

L: Let Go of FOMO

What are you afraid of missing? And what are you actually missing by staying glued to your devices?
- My biggest FOMO stories sound like: "If I don't check _____, then _____ will happen."
- What I actually tend to miss offline when I'm constantly checking
- Choose one FOMO story you're willing to question: "Maybe I don't have to chase this to be safe/relevant/loved."

U: Use Offline Alternatives

You can't just subtract. You have to swap. Some options: walk, stretch, doodle, text a friend, make tea, breathe, read a page, look out a window, pet an animal, lie on the floor for two minutes. For each major trigger, choose at least one offline alternative: When I feel _____, instead of scrolling _____, I can:

G: Generate Support

Who and what will help you keep going when motivation dips?
- People who "get it" (friends, family, colleagues)
- Professionals or groups (therapy, coaching, recovery, spiritual/meditation groups)
- Tools and reminders (printed planner, notes on the wall, alarms that say "time to unplug")
- Write one simple request for support you're willing to make: "Would you be willing to…?"

My Relapse Plan

When old patterns come back

You will slide back into old habits. That's not a moral failure; it's how brains work. What matters is what you do next. Use this page to write a plan for a future moment when you feel tired, overwhelmed, and tempted to give up.

When I notice I've slipped back into my old tech patterns

Instead of:
- Shaming myself
- Pretending it's not happening
- Throwing away the whole plan
- I will: Name it. "I'm in a loop with _____ again."

Feel it.
- What am I feeling right now?
- What do I actually need?

Do one small repair action. Examples:
- Turn off one app for the evening
- Put my phone in its "home" for an hour
- Go for a five-minute walk without headphones
- Write a few lines in this planner about what pulled me back in

Write a kinder story

Complete this sentence and come back to it:
"Relapse is not proof that I'm broken. It's a signal that _____."

Boredom Practice Menu

AKA Dopamine Fasting

These are practices you can keep returning to long after the 30 days are done. None of them are urgent. All of them are invitations. Pick one when you notice your hand reaching for a device "just because."

- **The 10-Minute Observation:** Look at one image, object, or view for ten minutes. Nothing else. Watch what your mind does.

- **Micro-Sketch:** Draw whatever is in front of you in 3 minutes: your mug, your hand, your keyboard, your cat. Bad art is welcome. It's about the process not the product.

- **Sensory Scan:** Sit still and list: 5 things you see, 4 you feel, 3 you hear, 2 you smell, 1 you taste.

- **Analog Brain Dump:** Open to a blank page and write down everything tugging at your attention. No organizing, just emptying.

- **Boredom Walk:** Walk without headphones or screens. Notice when boredom hits. Keep walking.

- **Pattern Hunt:** Look around and notice patterns: tiles, shadows, leaves, book spines, fabrics. List or sketch a few.

- **Slow Sips:** Drink something (coffee, tea, water) with full attention for five minutes. No multitasking.

- **One Song Only:** Put on one piece of music. Do nothing but listen – no scrolling, no email. When the song ends, pause before doing anything else.

- **Still-Frame Life:** Pick a moment (your desk, the sink, your bed) and imagine it's a movie still. What story would it tell?

- **Idea Jar:** Keep a small list or jar of tiny creative prompts (words, phrases, questions). When bored, pick one and spend five minutes playing with it.

- **Add your ideas below!**

How to Work with Me!

If you've made it to the end of this 30-day journey, you already know this work is not just about "screen time." It's about attention, nervous systems, creativity, grief, recovery, relationships, and how we stay human in a hyper-digital, AI-saturated world.

If these pages have sparked insights you want to take deeper into your own life, your organization, your classroom, or your community, the next step may be for us to work together directly.

Technology Detox & Digital Wellbeing Consulting

I offer consulting and advisory support for organizations, teams, and institutions who are ready to:

- Address technology overuse and burnout

- Design healthier digital cultures and norms

- Understand how UX, algorithms, and AI are shaping behavior

- Create policies and practices that protect attention, creativity, and wellbeing

Together we can audit current patterns, identify risks and opportunities, and build practical, humane strategies that work in the real world – not just on paper.

Speaking & Keynotes

I'm available for talks and keynotes on topics like:

- Technology Addiction & the Attention Economy

- Technology Detox in the Age of AI

- Boredom as a Practice for Creativity & Flow

- Embodied Leadership in the Age of AI

My talks are experiential, research-informed, and designed to leave audiences with inspiration, insights, and concrete next steps.

Workshops, Retreats & Trainings

For teams, campuses, and communities, I offer customized experiences including:

- Technology Detox workshops

- Boredom & Creativity labs

- Single-day and multi-day retreats and trainings based on the U.N.P.L.U.G. framework

These are highly interactive sessions that blend education, somatic practices, creative exercises, and supported behavior change.

How to Begin

I'd be honored to support you, your team, or your community in using less tech and living much more of your one wild, embodied, irreplaceable life.

Everything starts with a conversation. If you'd like to explore consulting, speaking, workshops, or retreat facilitation, reach out and tell me a bit about you, your context, and what you're hoping to change.

Let's talk: mindfulmarks.com/call
pattiebelle@mindfulmarks.com

Made in the USA
Middletown, DE
08 January 2026

26589093R00057